What Do You Know About
Fossils?

PowerKiDS
press
New York

Suzanne Slade

To the ever-exploring Boynton family—Bob, Linda, Mitch, Eric, and Mark

Published in 2008 by The Rosen Publishing Group, Inc.
29 East 21st Street, New York, NY 10010

First Edition

Editor: Amelie von Zumbusch
Book Design: Kate Laczynski
Photo Researcher: Nicole Pristash

Photo Credits: Cover, pp. 1, 5–9, 11–12, 16–17, 20 Shutterstock.com; p. 9 © Getty Images/Scott Olson; p. 10 © www.istockphoto.com/Sam Lee; p. 13 www.istockphoto.com/Falk Kienas; p. 14 © Getty Images/David McNew; p. 15 © AFP/Getty Images; p. 19 © Getty Images/Daniel LeClair; p. 21 © Getty Images; p. 22 © www.istockphoto.com/Tammy Bryngelson.

Library of Congress Cataloging-in-Publication Data

Slade, Suzanne.
  What do you know about fossils? / Suzanne Slade. — 1st ed.
     p. cm. — (20 questions : Science)
  Includes index.
  ISBN 978-1-4042-4197-8 (library binding)
  1. Dinosaur—Juvenile literature. 2. Fossils—Juvenile literature. I. Title.
  QE861.5.S55 2008
  560—dc22
                                                   2007023945

Manufactured in the United States of America

Contents

Fossils

Have you ever wondered what Earth was like **millions** of years ago? What interesting animals lived in the forest, flew through the air, and swam in the sea? Were there strange or beautiful plants?

Although we do not have pictures of Earth during that time, we do have fossils. Fossils are **preserved** animals and plants that lived long ago. These hard remains are an important record of Earth's history. Fossils give clues about early life on Earth. Many fossils are hidden in the earth. Maybe you will discover important secrets inside a fossil someday!

This sea turtle fossil shows us that the land where the fossil was found was once covered with a sea or ocean.

1. What's the big deal about fossils?

By studying fossils, **scientists** learn how Earth and living things have changed over time. Fossils show that some types of birds that lived millions of years ago were much larger than birds today. For example, the elephant bird was more than 6 feet (2 m) tall! Fossils also prove that certain animals, such as dinosaurs, have become **extinct**. Some fish fossils show that most of Earth was covered by water 500 million years ago. Animal fossils found in cold **Antarctica** prove this place was warm long ago.

Fossils like this one have shown scientists that crabs have been around for millions of years.

A scientist who studies fossils for a living is called a paleontologist, but anyone can search for fossils.

The paleontologists who uncovered this dinosaur fossil left it partly in the rock. This lets people see how the dinosaur's bones fit together.

3. Where can you find fossils?

Fossils are often found in sedimentary rock. This type of rock forms when **layers** of sand and mud harden over time.

4. Do fossil hunters use special tools?

When beginning a fossil dig, a **jackhammer** may be needed to break up rocks and hard dirt. Hammers and **chisels** also help loosen rocks. Fossil hunters move loose dirt with shovels. Scientists use a magnifying glass, or lens that makes things look bigger, to find tiny fossils. They use a

You can see several different layers of sedimentary rock in this cliff.

5. Are there different kinds of fossils?

The three main types of fossils are body fossils, imprint fossils, and trace fossils.

small tool like a dentist pick and soft brushes to clean fossils carefully.

Scientists use small, sharp tools to clean fossil teeth. These teeth came from a dinosaur called a Nanotyrannus.

6. What are body fossils?

Body fossils are parts of animals that have been preserved. Fossil eggs, teeth, and bones are all body fossils.

Dinosaur eggs, such as this one, are body fossils.

7. What is an imprint fossil?

Imprint fossils show scientists what ancient animals and plants looked like, although the actual remains are missing. The two kinds of imprint fossils are called molds and casts. A mold fossil is an empty space in a rock where a plant or animal used to be. A cast fossil is rock that has filled an empty mold fossil over time.

Scientists used fossil footprints, like the one here, to figure out that duck-billed dinosaurs often walked on two legs.

8. What are trace fossils?

A trace fossil is something an animal makes or a mark it leaves behind. A preserved animal tunnel, nest, or footprint are trace fossils.

This cast fossil was formed by a shell. If mud were to fill the cast fossil and harden into rock over time, it would form a mold fossil.

9. What kinds of fossils are made from plants?

Some trees in Arizona have changed into colorful fossils, called **petrified** wood. These fossils are about 220 million years old.

However, most plant fossils are imprint fossils. Scientists have discovered 400-million-year-old imprint fossils of **ferns**, mosses, and horsetails, bushy plants that look like a horse's tail.

If you visit Arizona's Petrified Forest National Park, you can see many logs of petrified wood.

10. Can bugs turn into fossils?

A few unlucky bugs were trapped in sticky tree sap long ago. After the sap hardened, it turned into amber. Amber is hard, smooth, and yellow. You can see through it like glass. Whole bug fossils, such as dragonflies and spiders, have been discovered inside amber. These discoveries lead scientists to believe that bugs first appeared on Earth about 365 million years ago.

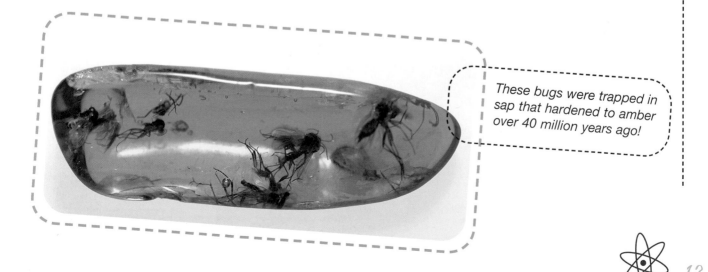

These bugs were trapped in sap that hardened to amber over 40 million years ago!

11. Do all living things become fossils?

Few living things end up as fossils. Most rot and change into dirt over time. Also, many plants and animals are eaten when they die. Fossils form when a plant or an animal is covered quickly by forces of nature, such as a flood or ash from a **volcano**.

These scientists are digging up a whale fossil. When the whale died, it sank to the ocean floor and was covered with mud, which later turned to rock.

Fossilized pieces of human **skeletons**, along with ancient human tools and footprints, have been discovered. These findings help scientists learn how people's bodies and lives have changed. In 2001, scientists discovered a five-million-year-old arm, hand, toe, and jaw, or mouth, bone in Africa. These are the oldest hominid, or humanlike, fossils found so far.

In 1994, scientists found the fossil of a hominid in South Africa's Sterkfontein caves. This fossil, which scientists named "Little Foot," is over three million years old.

13. Do fossils form in the ocean?

Cold water on the ocean floor helps keep the plants and animals there from rotting. Many turn into fossils over time. Fossils commonly found in the ocean are fish skeletons and teeth.

Fossils also commonly form in lakes. These fossils are of fish called knightia, which lived in a big lake that covered parts of Colorado, Utah, and Wyoming.

Ammonites, like this fossil, are one of the most common kinds of fossils. They died out millions of years ago but were related to today's squid and octopus.

14. Could I find any seashell fossils?

Around 400 million years ago, the sea was filled with snail-like animals called ammonites. Ammonites lived inside curved shells.

15. Are there any fossils trapped in ice?

About 20,000 years ago, most of Earth was covered with ice. This was called the Ice Age. Scientists have discovered **frozen** animals that look like furry elephants inside ice from this time. These animals are called woolly mammoths.

16. How do people know how old fossils are?

Scientists use several methods to find a fossil's age. The stratigraphy method uses how deep in the Earth a fossil is found to discover its age. This name comes from the word "strata," which means "layers of rock."

Scientists also study the rock around a fossil in the radiometric dating method. They test certain **elements** inside a rock that decay, or break down, over time. The test results date the rock and the fossil inside it.

17. How old are the oldest fossils?

The oldest fossils ever discovered are of tiny living things called algae that lived in the ocean. These fossils are three and a half **billion** years old!

In 2001, scientists discovered the fossil of an animal called a giant sloth. They used radiometric dating to figure out that the bones are eight and a half million years old.

18. Are there any really big fossils?

Dinosaur bones are among the largest fossils. The first dinosaur fossils were discovered in the early 1800s. Over the years, dinosaur bones, teeth, claws, and eggs have been unearthed. From these findings, scientists believe dinosaurs lived between 230 and 65 million years ago.

Though all the dinosaurs died out millions of years ago, scientists think that today's birds may be related to dinosaurs!

Dinosaur bones are large and heavy. Most dinosaur displays use lightweight copies of real dinosaur bones. These dinosaur skeletons are held together by pieces of steel. Wire keeps each bone in the right place. One of the most complete dinosaur skeletons was found in South Dakota by Sue Hendrickson in 1990. This *Tyrannosaurus rex* was later named Sue. You can visit it at Chicago's Field Museum.

People all over the world are interested in dinosaurs. This dinosaur display is at China's Nanjing Paleontology Museum.

20. How can I become a fossil hunter?

Fossils are hidden all over the world. Ask the parks department in your city or state if there are special fossil hunting places nearby. You can also look for fossils at the beach, in a **quarry**, or at the edge of a rocky cliff. Take an adult along when you hunt, as well as a map, shovel, bucket, and drinks. If you really dig fossils, you could become a paleontologist someday!

Looking for fossils is lots of fun. You never know what you might find!

Glossary

Antarctica (ant-AHRK-tih-kuh) The icy land at the southern end of Earth.

billion (BIL-yun) 1,000 millions.

chisels (CHIH-zulz) Sharp, metal tools used to cut and shape wood or stone.

elements (EH-luh-ments) The basic matter of which all things are made.

extinct (ek-STINKT) No longer existing.

ferns (FERNZ) Plants with big leaves called fronds.

frozen (FROH-zen) Hardened by great cold.

jackhammer (JAK-ha-mer) A tool for digging into rock.

layers (LAY-erz) Thicknesses of something.

millions (MIL-yunz) Thousands of thousands.

petrified (PEH-trih-fyd) Turned into stone after many years.

preserved (prih-ZURVD) Kept from being lost or from going bad.

quarry (KWOR-ee) A large hole, dug in the ground, from which stone is taken.

scientists (SY-un-tists) People who study the world.

skeletons (SKEH-leh-tunz) The things that give an animal's or a person's body shape.

volcano (vol-KAY-noh) An opening in Earth that sometimes shoots up hot, melted rock called lava.

Index

A
amber, 13
ammonites, 16
Antarctica, 6

D
dinosaur(s), 6, 20,
21

F
ferns, 12

H
Hendrickson, Sue,
21

M
mold(s), 10

P
paleontologist(s), 7,
22
petrified wood, 12

S
sedimentary rock, 8
skeleton(s), 15, 16,
21

V
volcano, 14

W
woolly mammoths,
17

Web Sites

Due to the changing nature of Internet links, PowerKids Press
has developed an online list of Web sites related to the subject of
this book. This site is updated regularly. Please use this link to
access the list:
www.powerkidslinks.com/20sci/foss/